ARMY RESEARCH LABORATORY

BRL-CAD Tutorial Series:
Volume I – Overview and Installation

by Lee A. Butler and Eric W. Edwards

ARL-SR-113

February 2002

Army Research Laboratory

Aberdeen Proving Ground, MD 21005-5068

ARL-SR-113   February 2002

BRL-CAD Tutorial Series: Volume I – Overview and Installation

Lee A. Butler
Survivability/Lethality Analysis Directorate, ARL

Eric W. Edwards
SURVICE Engineering Company

Abstract

Since 1979, the U.S. Army Research Laboratory has been developing and distributing the BRL-CAD constructive solid geometry (CSG) modeling package for a wide range of military and industrial applications. The package includes a large collection of tools and utilities including an interactive geometry editor, raytracing and generic framebuffer libraries, a network-distributed image-processing and signal-processing capability, and an embedded scripting language.

As part of this effort, a multivolume tutorial series is being developed to assist users in the many features of the BRL-CAD package. The "Overview and Installation" guide, which is the first volume in the series, addresses the background, purpose, and strengths of the package; the libraries and utilities included within it; platform-specific installation instructions; and information about bugs and updates. Other volumes focus on advanced features, individual utilities, and programming.

Acknowledgments

The authors would like to thank the members of the Advanced Computer Systems Team, who reviewed this document in a timely manner and made many helpful suggestions to improve its accuracy and presentation. Team members include John Anderson, TraNese Christy, Bob Parker, Ron Bowers, and Sean Morrison.

In addition, the authors would like to especially acknowledge Mike Muuss, a team member and the original architect of BRL-CAD, who passed away in the fall of 2000. Without his vision, this work would not have been possible. Therefore, the BRL-CAD Tutorial Series is dedicated to his memory.

INTENTIONALLY LEFT BLANK.

iv

Contents

INTENTIONALLY LEFT BLANK.

1. Introduction

1.1 What Is BRL-CAD?

Since the late 1950s, computers have been used to assist with the design and study of combat vehicle systems. The result has been a reduction in the amount of time and money required to take a system from the drawing board to full-scale production as well as increased efficiency in testing and evaluation.

In 1979, the U.S. Army Ballistic Research Laboratory (BRL) (now the U.S. Army Research Laboratory [ARL]) expressed a need for tools that could assist with the computer simulation and engineering analysis of combat vehicle systems and environments. When no existing computer-aided design (CAD) package was found to be adequate for this purpose, BRL software developers began assembling a suite of utilities capable of interactively displaying, editing, and interrogating geometric models. This suite became known as BRL-CAD.

Now comprising over one-half million lines of C code, BRL-CAD has become a powerful constructive solid geometry (CSG) modeling package that has been licensed at over 2,000 sites throughout the world. It contains a large collection of tools, utilities, and libraries including an interactive geometry editor, raytracing and generic framebuffer libraries, a network-distributed image-processing and signal-processing capability, and a customizable embedded scripting language. In addition, BRL-CAD simultaneously supports dual interaction methods, one using a command line and one using a graphical user interface (GUI).

A particular strength of the package lies in its ability to build and analyze realistic models of complex objects using a relatively small set of "primitive shapes." To do this, the shapes are manipulated by employing the basic Boolean operations of union, subtraction, and intersection. Another strength of the package is the speed of its raytracer, which is one of the fastest in existence. Finally, BRL-CAD users can accurately model objects on scales ranging from the subatomic through the galactic and get "all the details, all the time."

1.2 Why CSG Modeling?

Although BRL-CAD has been used for a wide variety of engineering and graphics applications, the package's primary purpose continues to be the support of (1) ballistic and (2) electromagnetic analyses. Accordingly, developers have found CSG modeling to be the best approach in terms of model accuracy, storage efficiency, precision, and speed of computational analysis.

While polygonal and boundary representation (B-rep) modeling often focuses on just the *surfaces* of objects, CSG modeling focuses on the entire *volume* and *content* of objects. This gives BRL-CAD the capability to be "more than skin deep" and build objects with real-world materials, densities, and thicknesses so that analysts can study physical phenomena such as ballistic penetration and thermal, radiative, neutron, and other types of transport.

2. Package Content

In keeping with the UNIX philosophy of developing independent tools to perform single, specific tasks and then linking the tools together in a package, BRL-CAD is basically a collection of libraries, tools, and utilities that work together to create, raytrace, and interrogate geometry and manipulate files and data. The basic data flow structure of the package is provided in Figure 1.

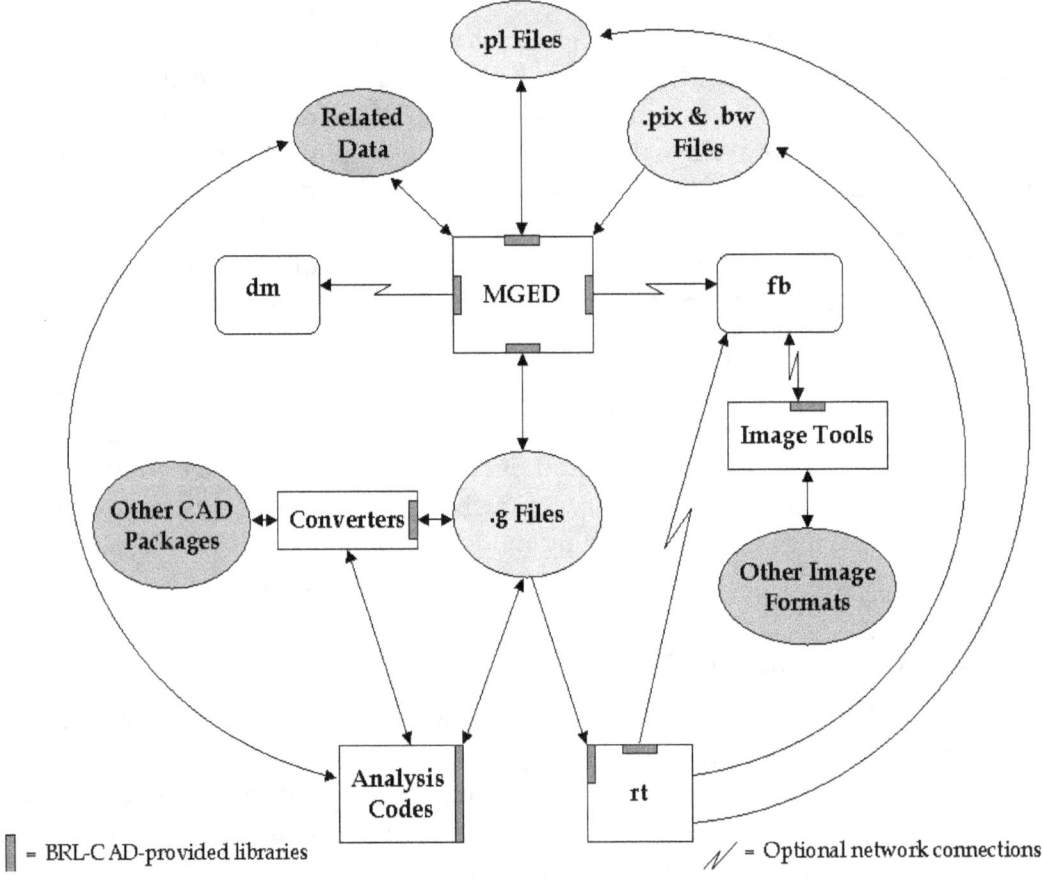

Figure 1. BRL-CAD data flow structure.

2.1 Libraries

The BRL-CAD libraries (designated by the prefix "lib") are designed primarily for the geometric modeler who also wants to tinker with software and, perhaps, design custom tools. Each library fits into one of three categories: (1) creating and/or editing geometry, (2) raytracing geometry, or (3) image handling. The following is a list of the major BRL-CAD libraries and descriptions of their functions.

- **libbu** – a basic utility (bu) library containing fundamental computer science types of routines, including manipulating data (e.g., converting from host format data to network format data), parallel processing, parsing parameters, handling variable-length strings, performing error checks, converting units from one format to another, manipulating bit vectors, running parallel-safe versions of input/output (I/O) routines, and maintaining symbol tables.

- **libbn** – a library of routines to support basic numerical (bn) handling, including 2-D/3-D vector, matrix, and quaternion manipulation; 3-D plotting support; automatic inference of image size (BRL-CAD images have no internal header; image dimensions are intuited from the size of the file); and wavelet decomposition and reconstruction, etc.

- **libdm** – BRL-CAD's primary graphics display manager (dm) library. It handles opening windows onto the display and displaying geometry in the graphics window, etc.

- **libfb** – the framebuffer (fb) library, which supports having a window in which the user puts pixel data when raytracing.

- **libmultispectral and liboptical** – the shader and texturing libraries for the raytracer.

- **libpkg** – a library that implements a remote procedure call (rpc) mechanism. This library is a predecessor of the modern rpc system. Unlike the typical UNIX rpc service, applications can set up services and handle requests without requiring configuration by a system administrator.

- **librt** – the library that contains all of the geometry support, including data representations for the primitives, support for raytracing (rt), and binary I/O support for CSG geometric descriptions.

- **libz** – a public-domain compression library.

- **libtcl, libtk,** and **libitcl** – libraries that provide the Tcl/Tk scripting language.

- **libpng** – a library that provides portable network graphics (png).

2.1 Tools and Utilities

The application side of BRL-CAD also offers a number of tools and utilities. They primarily concern (1) geometric conversion, (2) geometric interrogation, (3) image format conversion, and (4) command-line-oriented image manipulation. The following is a list of the major BRL-CAD tools and utilities.

- **MGED** (Multiple-Device Geometry Editor) – BRL-CAD's graphics editor. (For detailed guidance on the use of MGED as well as a list of all the MGED commands currently available, see Butler et al. [2001].)

- Tools for raytracing and interrogating raytraced geometric objects.

  - **rt** – the main raytracer for rendering images in BRL-CAD.

  - **nirt** – a package for firing rays interactively and getting information about what the rays run into.

  - **remrt** – a network-distributed raytracing package.

- An assortment of geometric converters to convert to and/or from other geometry formats, including Euclid, ACAD, AutoCAD DXF, TANKILL, Wavefront OBJ, Pro/ENGINEER, CATIA, JACK (the human factors model for doing workload/usability studies), Viewpoint Data Lab, NASTRAN, Digital Equipment's Object File Format (OFF), Virtual Reality Mark-up Language (VRML), Stereo Lithography (STL), Cyberware Digitizer data, and FASTGEN4.

- **bwish** – a Tcl/Tk interpreter in a windowing shell with enhancements compiled into it for accessing BRL-CAD libraries. It also includes various other extensions to the Tcl language.

- **irprep** – produces input to the PRISM (Physically Realistic Infrared Simulation Model) code.

- **JOVE** (Jonathan's Own Version of Emacs) – a fast, light implementation of Emacs.

- Applications for displaying images of various types on the framebuffer application and retrieving data from that framebuffer into images of various types.

- Tools for generating geometry for common objects such as fences, walls, and geometric mathematical oddities (e.g., the sphereflake shown in Figure 2 in Section 4).

- Data manipulation programs to (1) convert integers to floats, floats to doubles, etc. (e.g., **cv**); (2) perform mathematical operations on file elements (e.g., **imod, umod,** and **dmod**); (3) compute statistics of file elements (e.g., **istat, ustat,** and **dstat**); etc.

- Utilities for building animation scripts – keeping track of columnar data and interpolating it to allow one to produce input to the rt program to render multiple items for animation.

- **Utah Raster Tool Kit** – image manipulation of all RLE-based images.

- Programs for manipulating images and converting between different image file types. The two primary BRL-CAD types are **pix** (24-bit red, green, and blue [RGB] color images) and **bw** (8-bit greyscale images). Converters exist for various image formats including alias, png, ppm, etc.

- Programs for filtering images, doing histograms on the image data, and extracting rectangles from the images.

- Tools for combining two images and blending them together. (These tools were created before good image editing tools for video production were available; today users would typically load the images directly into a video editing package.)

---

3. Installation

3.1 Obtaining the Package

ARL currently offers two ways of obtaining the BRL-CAD package: (1) free distribution, and (2) full-service distribution. In both cases, a limited license agreement and survey (shown in the Appendix) must be signed by the user in order to receive the binary (executable) and/or source code. This agreement can be downloaded via anonymous file transfer protocol (FTP) from host <ftp.arl.army.mil>, file *brl-cad/agreement.pdf*, or via hypertext transfer protocol (http) from <http://ftp.arl.army.mil/brlcad/downloads/Current/agreement.pdf> (U.S. Army Research Laboratory 2001).

Free distribution (without support) is available from the aforementioned FTP site. Requests should be made to ARL at Aberdeen Proving Ground, MD, by way of fax (410-278-9177) or postal mail (BRL-CAD Distribution, ATTN: AMSRL-SL-B, APG, MD 21005-5068). After the license agreement is signed and processed, download instructions and the decryption key are returned to the requester. No further installation assistance or telephone assistance is available under the free distribution method.

Full-service distribution is available for a fee (or free to Government agencies) through the Aberdeen Satellite Office (ASO) of the Survivability Information Analysis Center (SURVIAC), which provides training and administers supported distributions and information exchange programs for ARL. The full-service distribution includes installation support; maintenance release updates; errata sheets; full technical support as needed via telephone, fax, e-mail, or postal mail; and information about future BRL-CAD activities. Requests should be made via telephone (410-273-7794), fax (410-272-6763), or postal mail (BRL-CAD Distribution, SURVIAC ASO, 4695 Millennium Drive, Belcamp, MD 21017-1505).

Any user who chooses the free distribution method to obtain BRL-CAD may upgrade to the full-service distribution at any time. In addition, all users have access to BRL-CAD symposia and workshops, the user's group, and the e-mail list (<cad@arl.army.mil>). Requests to join the list or questions regarding the list can be sent to <cad-request@arl.army.mil>.

3.2 System Requirements

Currently, BRL-CAD operates on a UNIX-based operating system and is supported on FreeBSD, RedHat Linux, SGI Irix, and Sun Microsystems Solaris. In addition, the package continues to be highly portable and has been run on other platforms, including Alliant FX/8, FX/80, and FX/2800; Apple Macintosh II running AUX; Convex C1; Cray-1,-2, X-MP, and Y-MP; Digital Equipment VAX; Gould/Encore PN 6000/9000; IBM RS/6000; Pyramid 9820; SGI 3030/Iris/Indigo/O2/Octane workstations; Sun-3 and Sun-4 SparcStation; and VAX/VMS.

Porting to other UNIX-based systems is also easily accomplished, usually with a day or two of effort. As of this writing, no port to Microsoft Windows has been completed.

3.3 Irix 6.x

The Irix binary distribution of BRL-CAD is intended for use on Irix 6.5. While the distribution may also work on older versions of Irix, support is limited. The "Irix6.5 32-bit installation tardist" was built using SGI's -n32 compile option. This package will work only on systems that have the -n32 libraries installed.

To install the file, perform the following steps:

(1) Download File

To download the 32-bit library version designed for Irix 6.5.distribution, shift-click on the link entitled "Irix6.x 32-bit installation tardist." Save the file to your local machine.

(2) Decrypt File

Use the decryption key (which was provided to you after you signed the license agreement) to decrypt the downloaded file. Type:

> *% crypt {key} < brlcad{version}_Irix65.tardist.crypt >*
> *brlcad{version}_Irix65.tardist*

> *NOTE:* The terms "*{key}*" and "*{version}*" throughout this document are not to be typed literally. They are used simply as placeholders for the appropriate decryption key value and BRL-CAD version number, respectively, which should be typed in place of them. Note also that there should be a space before and after "*{key}*" but not "*{version}*".

(3) Verify Download/Decryption

Now verify that the download and decryption were successful by running the following:

> *% md5 brlcad{version}**

Make sure that the resulting output matches that which is shown for the appropriate version on the web site.

(4) Install Package

Either use the "su"(switch user) command or log in as "root."

Remove any previous versions of BRL-CAD by typing the following:

> *# inst*
> *Inst> remove brlcad*.**
> *Inst> go*
> *Inst> quit*

Now install the new release with the following commands:

> *# inst -f brlcad{version}*.tardist*
> *Inst> install **
> *Inst> go*
> *Inst> quit*

When *inst* offers to save the distribution or remove it, select REMOVE. The only things that will be removed are the temporary files that *inst* unpacked from the tardist file.

The complete BRL-CAD distribution is now installed in /usr/brlcad; just add /usr/brlcad/bin to your search path, and you are ready to begin.

3.4 FreeBSD

To install the package on a FreeBSD platform, perform the following steps:

(1) Download File

Download the encrypted package by going to <http://ftp.arl.army.mil /brlcad/downloads/Current/freebsd.shtml> and shift-clicking on the link in the left side bar entitled "FreeBSD i86 package." Save the file (which is approximately 11 MB in size) to your local machine.

(2) Verify Crypt Program

Verify that the crypt program is operating correctly by running the following:

% echo testing | crypt g-123456 | od -c

The correct output is:

0000000 257 354   3 204 335 247 333 266

If instead you see the following:

0000000   h 242 025 237 260 335 256 351

then you have the MD5 version of libcrypt, rather than the DES version, installed. Alternatively, if you see the following:

0000000   i 211 314 336 J 265 H Z

then you have the correct version of libcrypt, but an older (broken) FreeBSD variant of the crypt(3) library routine.

Having Crypt Problems?

If you don't have the crypt command on your machine, or you get an error message about "makekey," or the test case doesn't decrypt, go to <http://ftp.arl.army.mil/brlcad/downloads/Current/freebsd_crypt.shtml> and, using the link on the left side bar, download the file *crypt_freebsd.tar.gz*. This tar file contains an executable for the enigma program. Unpack it in the same directory as the brlcad RPM file by running:

> *% tar zxvf enigma_freebsd.tar.gz*

Verify that the enigma program is operating correctly by running:

> *% echo testing | ./enigma g-123456 | od -c*

The correct output is:

> *WARNING: Character '-' is illegal in first two bytes of salt.*
> *Attempting to map in backwards-compatible manner.*
> *0000000 257 354 3 204 335 247 333 266*

Now decrypt your BRL-CAD distribution with the following command:

> *% ./enigma {key} < brlcad-{version}.tgz.crypt > brlcad-{version}.tgz*

This executable has the proper library routine already linked in and should function regardless of the library installed in /usr/lib. Now go back to the freebsd install page, verify the decryption, and continue the installation.

If the enigma program gives an error message or the md5 checksum of the decrypted file isn't correct, you may have the MD5 version of /usr/lib/libcrypt.* installed instead of the DES-compatible version. To test this, type the following command into a shell window:

> *% perl -e 'print crypt("glorp","gl"). "\n"'*

Now check the output. If it reads:

> *1gl$85n.KNIEKG0kYwxWxRO520*

you have the MD5 crypt() library, which is incompatible. If it reads:

> *gl4EsjmGvYQE*

you have the correct DES crypt() library.

WARNING: Be careful when correcting this problem. Installing the other library may cause problems with your /etc/passwd passwords. It's best to compile enigma from the source code (from <http://ftp.arl.army.mil/brlcad/downloads/Current/freebsd_crypt.shtml>) and forcibly link it against the proper library. If you don't understand how to do this, contact your system administrator for further assistance.

(3) Decrypt File

To decrypt the downloaded file, use the decryption key supplied to you when your license agreement was processed. Use the following command to decode the archive:

% crypt {key} < brlcad-{version}.tgz.crypt > brlcad-{version}.tgz

(4) Verify Download/Decryption

Run the following command to verify that the download and decryption were successful:

% md5 brlcad-{version}.tgz*

Make sure that the resulting output matches what is shown for the appropriate version on the web site.

(5) Install Package

Either use the "su" command or log in as "root."

If you have installed a previous version of BRL-CAD, you will need to uninstall it before installing the new version. To determine which version of BRL-CAD is installed on your system, type the following command:

pkg_info –a | egrep 'Information .* brlcad'

If BRL-CAD is installed, the previous command should produce output of the form:

Information for brlcad-{version}

To remove this package from your system, type the following:

pkg_delete brlcad-{version}

To install the new version of BRL-CAD, run the following command:

 # pkg_add brlcad-{version}.tgz

 *NOTE:* If you've not yet installed several optional system libraries, you may see the following message:

pkg_add: could not find package png-1.0.9 !

You can install the libraries via

pkg_add ftp://ftp7.freebsd.org/pub/FreeBSD/FreeBSD-current/packages /All/png-1.0.9.tgz

Alternatively, you can build the libraries from source, as follows:

cd /usr/ports/graphics/png; make install

The complete BRL-CAD distribution is now installed in /usr/brlcad; just add /usr/brlcad/bin to your search path, and you are ready to begin.

3.5 Linux

To install BRL-CAD on a Linux i86 platform, perform the following steps:

(1) Download File

Download the encrypted package by going to the <http://ftp.arl.army .mil/brlcad/downloads/Current/linux.shtml> web site and shift-clicking on the "Linux RPM" link in the left side bar. Save the file (which is approximately 12 MB in size) to your local machine.

(2) Decrypt File

To decrypt the downloaded file, use the decryption key that was provided to you after your license agreement was processed. Use the following command to decode the archive:

% crypt {key} < brlcad-{version}-0313.i386.rpm.crypt > brlcad-{version}-0313.i386.rpm

Having Crypt Problems?

If you don't have the crypt command on your machine, or you get an error message about "makekey," or the test case doesn't decrypt, go to <http://ftp.arl.army.mil/brlcad/downloads/Current/linux_crypt.shtml> and download the file *crypt_linux.tar.gz* using the link in the left side bar. This tar file contains an executable for the enigma program. Unpack it in the same directory as the brlcad RPM file by running the following:

> % *tar zxvf enigma_linux.tar.gz*

Verify that the enigma program is operating correctly by running the following:

> % *echo testing | ./enigma g-123456 | od -c*

The correct output is:

> *0000000 257 354   3 204 335 247 333 266*

Now decrypt your BRL-CAD distribution with the following command:

> % *./enigma {key} < brlcad-{version}-0302.i386.rpm.crypt >*
> *brlcad-{version}-0302.i386.rpm*

This executable has the proper library routine already linked in; it should function correctly regardless of which library you have installed in /usr/lib. Now go back to the Linux install page to verify the decryption.

If the enigma program gives you an error message or the md5 checksum of the decrypted file is not correct, you may have the MD5 version of /usr/lib/libcrypt.* installed instead of the DES-compatible version. To test this, type the following command into a shell window:

> % *perl -e 'print crypt("glorp","gl"). "\n"'*

Check the output in the following table. If it reads

> *1gl$85n.KNIEKG0kYwxWxRO520*

you have the MD5 crypt() library, which is incompatible. If it reads

> *gl4EsjmGvYQE*

you have the correct DES crypt() library.

WARNING: Be careful when correcting this problem because installing the other library may cause problems with your /etc/passwd passwords. It's best to compile enigma from the source code (from <http://ftp.arl.army.mil/brlcad/downloads/Current/linux_crypt.shtml>) and forcibly link it against the proper library. If you don't know how to do this, contact your system administrator for further assistance.

(3) Verify Download/Decryption

Verify that the download and decryption were successful by running the following:

> *% md5sum brlcad-{version}**

Make sure that the resulting output matches that shown for the appropriate version on the web site.

(4) Install Package

Either use the "su" command or log in as "root."

Run the following command:

> *# rpm -ihv brlcad-{version}-0313.i386.rpm*

If you have installed a previous version of BRL-CAD, you will need to specify "upgrade" rather than "install." You can do this with the following command:

> *# rpm -Uhv brlcad-{version}-0313.i386.rpm*

If you have an old non-RPM BRL-CAD in /usr/brlcad, do an "rm -rf /usr/brlcad" as root before starting the installation.

The complete BRL-CAD distribution is installed in /usr/brlcad; just add /usr/brlcad/bin to your search path, and you are ready to begin.

3.6 Solaris

To install BRL-CAD on a Solaris SPARC platform, perform the following steps:

(1) Download File

Download the encrypted package by going to the <http://ftp.arl.army.mil/brlcad/downloads/Current/solaris.shtml> web site and shift-clicking on the link entitled "Solaris package" on the left side bar. Save the file (which is approximately 20 MB in size) to your local machine.

(2) Verify Crypt Program

Verify that the crypt program is operating correctly by running the following:

> *% echo testing | crypt g-123456 | od -c*

The correct output is:

> *0000000 257 354 3 204 335 247 333 266*
> *0000010*

(3) Decrypt File

To decrypt the downloaded file, use the decryption key that was supplied to you after your licensing agreement was processed. Use the following command to decode the archive:

% crypt {key} < brlcad-{version}-2001.03.13-solaris.gz.crypt | gunzip > brlcad-{version}-2001.03.13-solaris

(4) Verify Download/Decryption

Verify that the download and decryption were successful by running the following:

% md5 brlcad-{version}-*solaris*

Make sure that the resulting output matches that shown for the appropriate version on the web site.

(5) Install Package

Either use the "su" command or log in as "root."

If you have installed a previous version of BRL-CAD, you will need to uninstall it before installing the new version. To do this, run the following command:

 # pkgrm brlcad

To install the new version of BRL-CAD, run the following command:

 # pkgadd brlcad-{version}-2001.03.13-solaris

The complete BRL-CAD distribution is now installed in /usr/brlcad; just add /usr/brlcad/bin to your search path, and you are ready to begin.

3.7 Source Code

Performing an installation from source involves more effort than a binary distribution. Users are encouraged to install the appropriate binary distribution whenever possible. However, for special requirements or for systems in which a binary distribution is not available, the source distribution is the only way to proceed. To install from source, perform the following steps:

(1) Download File

Download the encrypted file by going to the <http://ftp.arl. army.mil/brlcad/downloads/Current/src.shtml> web site and shift-clicking on link entitled "Source archive" on the left side bar. Save the gzipped "tar" file (which is approximately 17 MB in size) to your local machine somewhere other than the directory where you intend to install the BRL-CAD products.

 *NOTE:* If you have problems downloading this large file, you can download each file individually by selecting them, in turn, under the "Smaller Downloads" links on the left side bar of the <http://ftp.arl .army.mil/brlcad/downloads/Current/src.shtml> web site.

(2) Verify Download

To verify that the download was successful on each file, run the following command(s) (On a Linux platform, you may need to run md5sum instead.):

> *% md5 cad{version}.tar.gz.crypt*

or

> *% md5 cad{version}.tar-?.gz.crypt*

(3) Decrypt File

To decrypt the downloaded file, use the decryption key provided to you after processing of your license agreement (or after receiving a prior release). Use the following command to decode the archive:

> *% crypt {key} < cad{version}.tar.gz.crypt > cad5.3.tar.gz*

If you downloaded the smaller files instead, you must do this to each file, in turn, as follows:

> *% crypt {key} < cad{version}.tar-a.gz.crypt > cad{version}.tar-a.tar.gz*
> *% crypt {key} < cad{version}.tar-b.gz.crypt > cad{version}.tar-b.tar.gz*
> etc.

(4) Uncompress/Unpack Archive

If you downloaded the entire file at once, run the following command:

> *% gunzip < cad{version}.tar.gz | tar xf –*

If you downloaded the smaller files instead, run the following command for each file, in turn:

> *% gunzip < cad{version}.tar-a.tar.gz | tar xf -*
> *% gunzip < cad{version}.tar-b.tar.gz | tar xf –*
> etc.

Be sure the destination directory for the installation exists. Type

> *% mkdir /usr/brlcad*

If you choose a directory other than /usr/brlcad, you will need to set the environment variable BRLCAD_ROOT to be the name of this directory.

A number of directories are created under the root directory for the software package. The full names of these are compiled into various parts of BRL-CAD. As a result, it is not possible to compile the package for one location and later relocate the binaries to another.

Be sure that /usr/brlcad/bin or $BRLCAD_ROOT/bin is in your shell path as follows. For /bin/sh:

> *BRLCAD_ROOT=/usr/brlcad*
> *PATH=$PATH:/usr/brlcad/bin*
> *export PATH*

Likewise, for csh:

> *setenv BRLCAD_ROOT /usr/brlcad*
> *set path=($path /usr/brlcad/bin)*

(5) Compile Package

Now compile the package as follows:

> *% cd brlcad*
> *% sh setup.sh*

The next step can take a considerable amount of time on some systems. For sh, type the following:

> *% make > make.log 2>1*

For csh, type the following:

> *% make >& make.log*

(6) Install Package

Before installing the package, review the log of compilation for errors. If there are no errors, then type the following:

> *% make install*

The complete BRL-CAD distribution is now installed in /usr/brlcad; just add /usr/brlcad/bin to your search path, and you are ready to begin.

4. Benchmark Testing

As changes are implemented in BRL-CAD, ARL developers run a standard set of computationally intensive image files (shown in Figure 2) on a common machine in order to benchmark and compare raytrace performance. In addition, these images are provided with each source distribution of the package so that users can also test performance on their machines, if desired.

To run the benchmark images, run the script *run.sh* in the "bench" directory of the source directory tree.

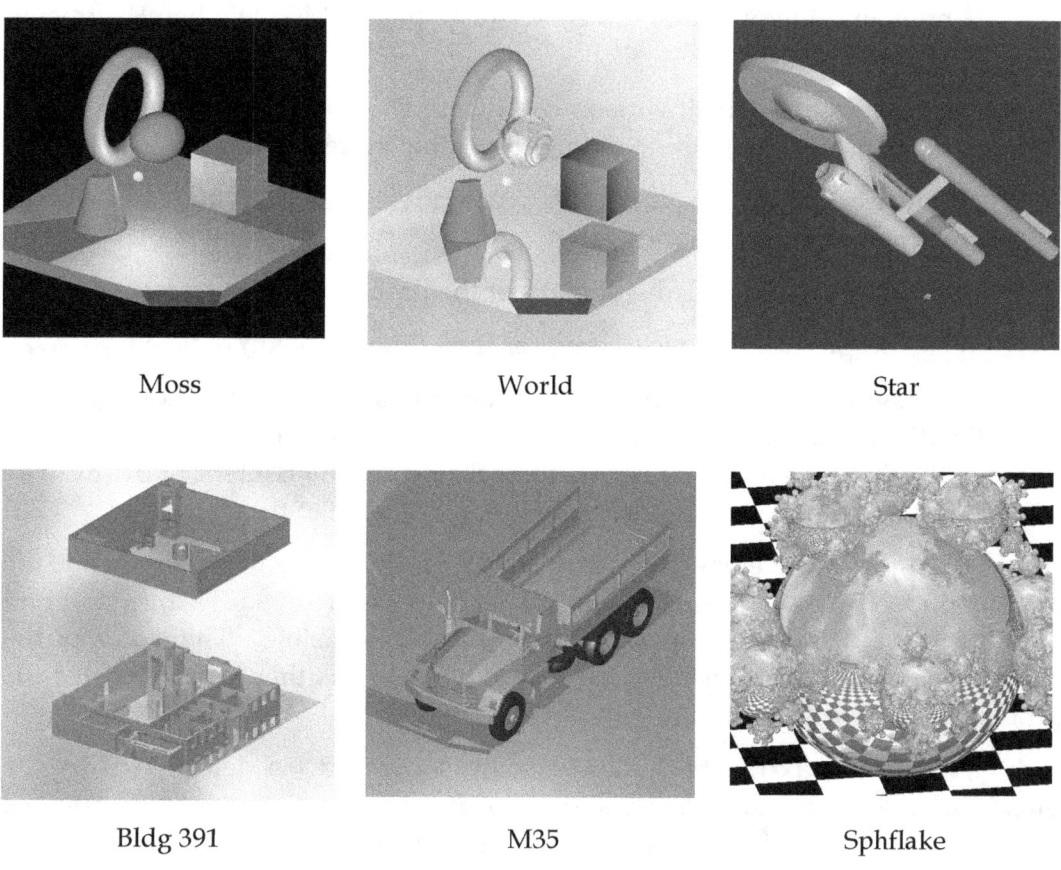

Moss   World   Star

Bldg 391   M35   Sphflake

Figure 2. Benchmarking images.

5. Maintenance and Updates

5.1 Troubleshooting and Tracking Bugs

When a bug or deficiency is found in BRL-CAD, the user should report the problem to ARL developers as soon as possible. This is easily accomplished by filing a Bug Report via the shell script *cad-bug.sh* (which is provided with each distribution of the package). In order to process the bug quickly and efficiently, the following information should be included in the Bug Report:

- **Abstract** – a short (40 characters or fewer) tag line indicating the name of library and the problem/deficiency (e.g., Bug in MGED when Raytracing Spheres).

- **Environment** – which version of BRL-CAD is being used, the hardware on which it is being used, and the operating system.

- **Repeat-By** – the sequence of procedures that preceded/caused the problem.

- **Description** – a detailed summary of the problem or suggestion.

- **Fix** – proposed solutions (if any) for fixing the problem/deficiency.

BRL-CAD bugs are filed into one of three categories in the Bug Report Registry: New, Pending, and Closed. When notice of a bug is received, the report is filed in the New directory. Due to regulations regarding the content of Army web sites, new messages are not immediately made available to the public.

After a BRL-CAD developer has verified that a new Bug Report is valid, the aforementioned script message is automatically converted to HTML format and placed into the Pending directory, which is publicly visible. The status of the message is then continuously updated by developers as they work to solve the problem.

Finally, when the fix has been satisfactorily addressed (as determined by the BRL-CAD developer), the report and all of its amendments are moved to the Closed directory, where they remain in a publicly viewable archive.

5.2 New Releases

As new versions of BRL-CAD are released, announcements and release notes are made through the <cad@arl.army.mil> list. Interested users can then download the encrypted files from the Internet and decrypt them using their original decryption key.

6. References

Butler, Lee A., Eric W. Edwards, Betty J. Schueler, Robert G. Parker, and John R. Anderson. "BRL-CAD Tutorial Series: Volume II - Introduction to MGED." ARL-SR-102, U.S. Army Research Laboratory, Aberdeen Proving Ground, MD, April 2001.

U.S. Army Research Laboratory. <http://ftp.arl.army.mil/brlcad/> and subpages, Aberdeen Proving Ground, June 2001.

INTENTIONALLY LEFT BLANK.

Appendix: BRL-CAD License Agreement

License Agreement
Statement of Terms and Conditions for Release of
The BRL-CAD Package

The Federal Government of the United States of America is hereinafter "THE GOVERNMENT".

The U. S. Army Research Laboratory, hereinafter "ARL", an agency of THE GOVERNMENT, is the originator of The BRL-CAD Package.

The BRL-CAD Package is a collection of computer programs, data files, and associated documentation, hereinafter "BRL-CAD". BRL-CAD is provided in machine-readable form and may be obtained in either "binary executable" or "source code" editions. This license applies to both editions.

BRL-CAD and the "BRL-CAD Eagle" are trademarks of the United States Army.

The DTIC Survivability/Vulnerability Information Analysis Center, hereinafter "SURVIAC", administers the supported BRL-CAD distributions and information exchange programs on behalf of ARL.

The corporation or private individual requesting BRL-CAD as identified on page two is hereinafter the "RECIPIENT".

Changes made to any files provided with BRL-CAD are hereinafter "MODIFICATIONS". New files entirely created by RECIPIENT are hereinafter "EXTENSIONS".

1. BRL-CAD is an unpublished work that is not generally available to the public, except through the terms of this limited distribution. ARL grants RECIPIENT a personal, non-exclusive, non-transferable license and right to use BRL-CAD. No right is granted for any use of BRL-CAD by any third party.

2. RECIPIENT will be responsible for assuring that BRL-CAD will not be released or sold to any other party without the prior written approval of ARL.

3. RECIPIENT guarantees that BRL-CAD, and any modified versions thereof, will not be published for profit or in any manner offered for sale to THE GOVERNMENT. BRL-CAD may be used in contracts with THE GOVERNMENT but no development charge may be made as part of its use.

4. If RECIPIENT makes MODIFICATIONS to BRL-CAD, then RECIPIENT agrees to provide a copy of all such MODIFICATIONS to ARL in source code form. RECIPIENT agrees that such MODIFICATIONS may be used by ARL and the entire BRL-CAD user community.

5. RECIPIENT will own full rights to any EXTENSIONS to BRL-CAD they create.

6. RECIPIENT will own full rights to any data files, databases or images created using BRL-CAD.

7. BRL-CAD is provided "as is", without warranty. Neither THE GOVERNMENT nor SURVIAC is liable or responsible for maintenance, updating, or correcting any errors in any materials provided. In no event shall THE GOVERNMENT or SURVIAC be liable for any loss or for any indirect, special, punitive, exemplary, incidental, or consequential damages arising from use, possession, or performance of BRL-CAD.

($Revision: 1.28 $)

8. The RECIPIENT shall indemnify and hold harmless THE GOVERNMENT for any loss, claim, damage, expense, or liability of any kind occurring as a result of the making, using, or selling of a product, process, or service by or on behalf of RECIPIENT, its assignees and licensees, which was derived from BRL-CAD.

9. ARL will be credited should BRL-CAD be used in a product or written about in any publication. ARL will be referenced as the original source for BRL-CAD, however, the RECIPIENT shall not in any way imply that THE GOVERNMENT endorses any product or service of the RECIPIENT.

10. RECIPIENT agrees that when bugs or problems are found with BRL-CAD, a reasonable attempt will be made to report them to SURVIAC or ARL.

11. RECIPIENT agrees to complete and return the "Recipient Survey Form" attached below.

By signing here, RECIPIENT signifies agreement to these terms and conditions of the BRL-CAD distribution license agreement as detailed above.

Recipient's Signature

Recipient's Printed Name

Title

Corporation

Mailing Address

City, State, Zip Code, Country

Phone #

FAX #

Internet E-mail Address

Date

($Revision: 1.28 $)

Recipient Survey Form for
The BRL-CAD Package

The information collected on this form will be used to support the BRL-CAD effort, and will be available only to appropriate ARL and SURVIAC personnel unless indicated otherwise, below. If RECIPIENT grants permission to release this information, ARL and SURVIAC will make it available to BRL-CAD related hardware vendors and service providers.

1. [ ] RECIPIENT grants ARL and SURVIAC permission to release information on this "Recipient Survey Form" to outside parties.
   [ ] RECIPIENT considers this information to be proprietary and does not grant permission to release this information to outside parties.

2. BRL-CAD is being obtained: [ ] free without support privileges   [ ] with full-service support.

3. Please designate a technical point-of-contact:
   Mark here [ ] and leave this section blank if same as RECIPIENT on page 2.

Printed Name
Title
Corporation
Mailing Address
City, State, Zip Code, Country
Phone #
FAX #
Internet E-mail Address

4. Please indicate the type of your organization:
   [ ] Government: [ ] Air Force   [ ] Army   [ ] Navy   [ ] Other _____
   [ ] Industry ....(Government contractors should register as Industry).
   [ ] Academic .... (Students should register as Individuals).
   [ ] Private Individual

5. Number of people in work group _____

6. Have you used BRL-CAD before? [ ] Yes   [ ] No   If yes, for how many years? _____

7. Are you interested in participating in the BRL-CAD user group? _____

($Revision: 1.28 $)

8. Please indicate the anticipated uses of BRL-CAD at this site (check all that are appropriate):

[ ] Computer Aided Design (CAD)   [ ] Animation   [ ] Image Processing
[ ] Computer Aided Manufacturing (CAM)   [ ] Visualization   [ ] Education
[ ] Vulnerability Assessment   [ ] Architectural Design   [ ] Graphic Arts
[ ] Signature Prediction and Analysis   [ ] Other (please elaborate)   _____

9. Please list the computer platforms that you are likely to install BRL-CAD on (check all that are appropriate):

Vendor	CPU Model	OS Revision
[ ] DEC		
[ ] IBM		
[ ] Linux		
[ ] FreeBSD		
[ ] Sun		
[ ] SGI		
[ ] Other (specify)		

10. Types of BRL-CAD users at this site (check all that are appropriate):

[ ] Management   [ ] Draftsperson   [ ] Graphic artist
[ ] Engineer   [ ] Designer   [ ] Image Processor   [ ] Other _____

11. Anticipated use of BRL-CAD by at least one person at this site:

[ ] >10 days/month   [ ] 5-10 days/month
[ ] 1-4 days/month   [ ] several times/year

12. Please specify which edition of BRL-CAD you desire:
    [ ] 5.0 Binary executable edition (versions exist for SGI-IRIX-6, RedHat Linux-86, and FreeBSD-86 architectures).
    [ ] 5.0 Source code edition (please answer question 13 below).
    [ ] 4.5 Source code edition.

13. If you desire the source code edition, please indicate briefly your reasons and the ways you envision using the source code:

_____

_____

_____

14. If you are purchasing physical media from SURVIAC, please specify the format:

[ ] No media, using free network FTP access
[ ] CD-ROM
[ ] Zip(tm) 100 MByte cartridge, MS-DOS format

($Revision: 1.28 $)

Instructions for completing the form

For record-keeping purposes, the five (5) pages of the completed and signed form must be returned either (a) printed on paper, or (b) transmitted via facsimile machine (FAX).

Free Distribution

If in item 2 you requested a "FREE distribution with no support privileges" from ARL for FTP file transfer over the Internet, please transmit the completed form via FAX to USA telephone +1.410.278.9177, or send via postal mail to:

> BRL-CAD Distribution
> Attn: AMSRL-SL-B
> APG, MD 21005-5068 USA

Download instructions and the decryption key will be returned to you by FAX or US Mail. ARL and SURVIAC regret that they cannot accept inquiries about the free distribution by telephone.

Full-Service Distribution

If you are purchasing a "Full-service distribution" from SURVIAC, please complete and return a signed copy of the distribution agreement and survey form with a check or purchase order for $500 payable to "BA&H/SURVIAC" via FAX to USA 410-272-6763, or send via postal mail to:

> BRL-CAD Distribution
> SURVIAC Aberdeen Satellite Office
> 4695 Millenium Drive
> Belcamp, MD 21017-1505 USA

Included with this full-service distribution are installation support, maintenance release updates, technical support, and information about future activities. For further details, contact "BRL-CAD full-service distribution" at USA 410-273-7794 or send e-mail to cad-dist@arl.mil.

($Revision: 1.28 $)

INTENTIONALLY LEFT BLANK.

NO. OF COPIES	ORGANIZATION	NO. OF COPIES	ORGANIZATION
1	DEFENSE TECHNICAL INFORMATION CENTER DTIC OCA 8725 JOHN J KINGMAN RD STE 0944 FT BELVOIR VA 22060-6218	1	AIR FORCE ARMAMENT CNTR AAC/ENMA D MCCOWN 101 W EGLIN BLVD EGLIN AFB FL 32542-5549
1	DIRECTOR US ARMY RESEARCH LAB AMSRL D R W WHALIN 2800 POWDER MILL RD ADELPHI MD 20783-1197	1	USAF 46 OG OGMLV B THORN 104 CHEROKEE AVE EGLIN AFB FL 32542-5600
1	DIRECTOR US ARMY RESEARCH LAB AMSRL D D SMITH 2800 POWDER MILL RD ADELPHI MD 20783-1197	1	USAF WRIGHT LABORATORY 46TH OG OGM AL AC M LENTZ 2700 D STREET BLDG 22B WRIGHT PAT AFB OH 45433-7605
3	DIRECTOR US ARMY RESEARCH LAB AMSRL CI LL 2800 POWDER MILL RD ADELPHI MD 20783-1197	1	SURVIAC ABERDEEN SATELLITE OFC A LAGRANGE 4695 MILLENNIUM DRIVE BELCAMP MD 21017-1505
1	DIRECTOR US ARMY RESEARCH LAB AMSRL CI AI R 2800 POWDER MILL RD ADELPHI MD 20783-1197	6	THE SURVICE ENGNRG CO D KREGEL B STRAUSSER C BOYER M HARDIN M BUTKIEWICZ L MCKAY 4695 MILLENNIUM DRIVE BELCAMP MD 21017-1505
1	DIRECTOR US ARMY RESEARCH LAB AMSRL CI AP 2800 POWDER MILL RD ADELPHI MD 20783-1197		
1	DIRECTOR US ARMY RESEARCH LAB AMSRL SL C HOPPER WSMR NM 88002-5513		ABERDEEN PROVING GROUND
		4	DIR USARL AMSRL CI LP (305)
1	NAWC WEAPONS DIVISION CODE 418300D A WEARNER BLDG 91073 1 ADMINISTRATION CIRCLE CHINA LAKE CA 93555-6100	220	DIR AMSRL AMSRL SL DR WADE J BEILFUSS AMSRL SL E M STARKS AMSRL SL EC E PANUSKA AMSRL SL EM J FEENEY

27

NO. OF
COPIES   ORGANIZATION

AMSRL SL B (5 CPS)
AMSRL SL BA (25 CPS)
AMSRL SL BD (15 CPS)
AMSRL SL BE (25 CPS)
    L BUTLER (100 CPS)
AMSRL SL BG (25 CPS)
AMSRL SL BN (20 CPS)

REPORT DOCUMENTATION PAGE

Form Approved
OMB No. 0704-0188

Public reporting burden for this collection of information is estimated to average 1 hour per response, including the time for reviewing instructions, searching existing data sources, gathering and maintaining the data needed, and completing and reviewing the collection of information. Send comments regarding this burden estimate or any other aspect of this collection of information, including suggestions for reducing this burden, to Washington Headquarters Services, Directorate for Information Operations and Reports, 1215 Jefferson Davis Highway, Suite 1204, Arlington, VA 22202-4302, and to the Office of Management and Budget, Paperwork Reduction Project(0704-0188), Washington, DC 20503.

1. AGENCY USE ONLY *(Leave blank)*	2. REPORT DATE February 2002	3. REPORT TYPE AND DATES COVERED Final, May 2001 - July 2001

4. TITLE AND SUBTITLE BRL-CAD Tutorial Series: Volume I - Overview and Installation	5. FUNDING NUMBERS 1L162618AH80
6. AUTHOR(S) Lee A. Butler and Eric W. Edwards*	

7. PERFORMING ORGANIZATION NAME(S) AND ADDRESS(ES) U.S. Army Research Laboratory ATTN: AMSRL-SL-BE Aberdeen Proving Ground, MD 21005-5068	8. PERFORMING ORGANIZATION REPORT NUMBER ARL-SR-113

9. SPONSORING/MONITORING AGENCY NAMES(S) AND ADDRESS(ES)	10. SPONSORING/MONITORING AGENCY REPORT NUMBER

11. SUPPLEMENTARY NOTES

*Mr. Edwards is employed by the SURVICE Engineering Company, headquartered in Belcamp, MD.

12a. DISTRIBUTION/AVAILABILITY STATEMENT Approved for public release; distribution is unlimited.	12b. DISTRIBUTION CODE

13. ABSTRACT *(Maximum 200 words)*

Since 1979, the U.S. Army Research Laboratory has been developing and distributing the BRL-CAD constructive solid geometry (CSG) modeling package for a wide range of military and industrial applications. The package includes a large collection of tools and utilities, including an interactive geometry editor, raytracing and generic framebuffer libraries, network-distributed image-processing and signal-processing capabilities, and an embedded scripting language.

As part of this effort, a mulivolume tutorial series is being developed to assist users in the many features of the BRL-CAD package. The "Overview and Installation" guide, which is the first volume in the series, addresses the background, purpose, and strengths of the package; the libraries and utilities included within it; platform-specific installation instructions; and information about bugs and updates. Other volumes focus on advanced features, individual utilities, and programming.

14. SUBJECT TERMS BRL-CAD, computer-aided design, solid modeling, CSG, software installation	15. NUMBER OF PAGES 30
	16. PRICE CODE

17. SECURITY CLASSIFICATION OF REPORT UNCLASSIFIED	18. SECURITY CLASSIFICATION OF THIS PAGE UNCLASSIFIED	19. SECURITY CLASSIFICATION OF ABSTRACT UNCLASSIFIED	20. LIMITATION OF ABSTRACT UL

INTENTIONALLY LEFT BLANK.

www.ingramcontent.com/pod-product-compliance
Lightning Source LLC
Chambersburg PA
CBHW081406170526
45166CB00010B/3235